MY FIRST
P

GREAT MAMMALS

Consultant: Ronald M. Nowak
Illustrators: Barbara Gibson, Tim Phelps

Published by
The National Geographic Society
John M. Fahey, Jr., President and Chief Executive Officer
Gilbert M. Grosvenor, Chairman of the Board
Nina D. Hoffman, Senior Vice President
William R. Gray, Vice President and Director, Book Division

Staff for this Book
Barbara Lalicki, Director of Children's Publishing
Barbara Brownell, Senior Editor and Project Manager
Marianne R. Koszorus, Senior Art Director and Project Manager
Toni Eugene, Editor
Alexandra Littlehales, Art Director
Carolinda Hill, Writer-Researcher
Susan V. Kelly, Illustrations Editor
Carl Mehler, Senior Map Editor
Jennifer Emmett, Assistant Editor
Mark A. Caraluzzi, Director of Direct Response Marketing
Vincent P. Ryan, Manufacturing Manager
Lewis R. Bassford, Production Project Manager

Visit our Web site at www.nationalgeographic.com

Library of Congress Catalog Card Number: 97-76351
ISBN: 0-7922-3452-9

Color separations by Quad Graphics, Martinsburg, West Virginia
Printed in Mexico by R.R. Donnelly & Sons Company

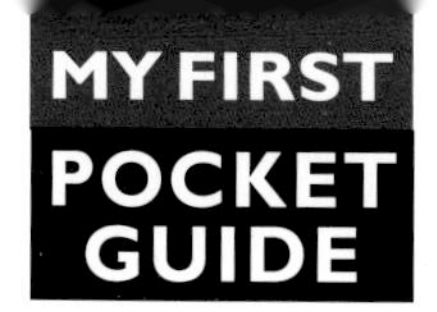

GREAT MAMMALS

CAROLINDA HILL

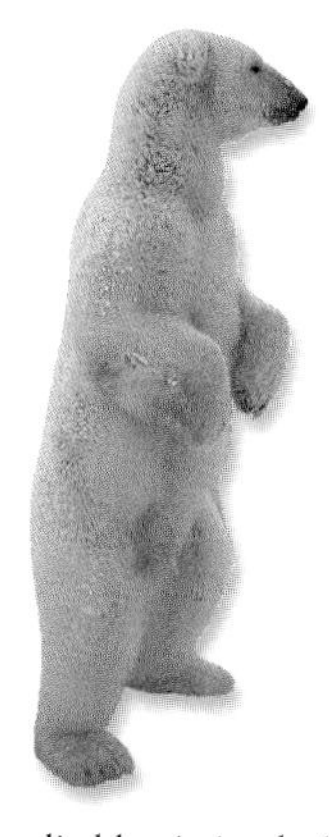

All photographs supplied by Animals Animals/Earth Scenes

NATIONAL
GEOGRAPHIC
SOCIETY

INTRODUCTION

Like all other mammals, great mammals are warm-blooded, which means they keep a constant body temperature even when the temperature of their surroundings varies. Like all mammals, great mammals feed milk to their young and have some hair or fur. What sets great mammals apart is their size: They are big. One of them—the blue whale—is not only the largest animal in the world but also the biggest creature that has ever lived.

Great Mammals looks at the heaviest and tallest land and sea animals, as well as other mammals that are the largest of their kind, including the greatest ape, the biggest cat, and the most humongous wild hog. Many of these extraordinary creatures live in North America. You can find them in wilderness areas, parks, and reserves. Some swim off the coast, and you might spot them from shore. Your local zoo is also home to many large mammals, including ones from other continents. Sadly, about half the animals in this book are threatened with extinction.

HOW TO USE THIS BOOK

This book is organized by size—from the largest great mammal to the smallest. Each spread helps you identify 1 of 35 kinds of animals and tells you about its size, color, and behavior. A paw or hoof print shows the track made by the land animal. A shaded map of the world indicates where the mammal lives in the wild, and the "Field Notes" entry gives an unusual fact about it. If you come across a word you do not know, look it up in the Glossary on page 76.

BLUE WHALE

The blue whale is the largest animal in the world. In summer, it eats four to eight tons of tiny shrimp-like creatures each day. For the rest of the year, it may eat nothing at all, living on stored fat.

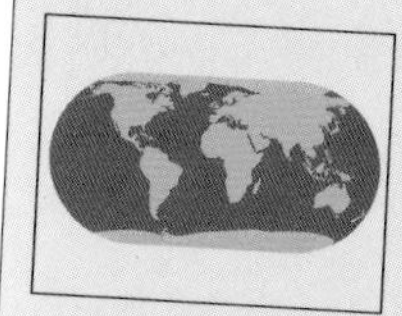

WHERE TO FIND:

Blue whales swim in every ocean, but their small numbers make these great mammals hard to find.

WHAT TO LOOK FOR:

✳ SIZE
Blue whales can weigh more than 200 tons and grow 100 feet long.

✳ COLOR
These whales are bluish gray, with white or light-yellow bellies.

✳ BEHAVIOR
A blue whale swims alone or with two or three other whales.

✳ MORE
The ear openings of a blue whale are about the size of the lead in a pencil.

Two-foot-thick body fat, called blubber, helps keep all whales warm in cool seas.

FIELD NOTES

This biggest whale is longer than five elephants walking in single file, and it is six times as heavy.

ORCA

World's largest dolphin (DOLL-fin), or small toothed whale, the orca cruises near coastlines. Its shark-like fin slices the water. People call orcas killer whales because they eat sea mammals.

FIELD NOTES

Looking for a meal, an orca heaves itself onto shore and snaps up a sea lion, which it swallows whole.

Orcas whistle, scream, and make clicking sounds to talk to each other.

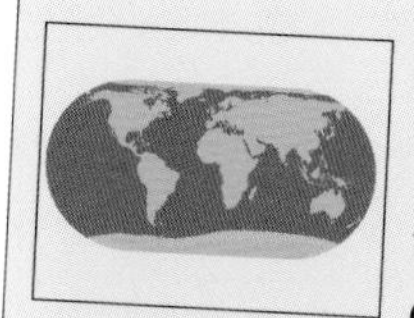

WHERE TO FIND:

Orcas live in every ocean and in nearly all of the world's seas. They prefer cool coastal waters.

WHAT TO LOOK FOR:

✱ SIZE
An orca can reach 32 feet in length and weigh as much as 10 tons.

✱ COLOR
It is black, with white on the jaws, belly, and sides of the head.

✱ BEHAVIOR
Five to forty killer whales travel together in groups called packs.

✱ MORE
Orcas can swim faster than 25 mph and can stay underwater almost 20 minutes.

AFRICAN ELEPHANT

Biggest of all land animals, an African elephant outweighs elephants living in Asia. Its tusks, which weigh up to 200 pounds each, are the world's largest teeth. Its six-foot-long trunk is the largest nose.

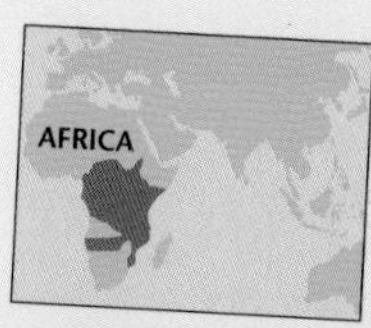

WHERE TO FIND:
African elephants roam grasslands, woodlands, and river valleys. They usually avoid desert areas.

WHAT TO LOOK FOR:

✳ SIZE
African elephants weigh more than 6 tons and stand 13 feet tall.

✳ COLOR
Elephants are usually brownish gray.

✳ BEHAVIOR
They may eat 500 pounds of food and drink 60 gallons of water a day.

✳ MORE
They can make sounds humans cannot hear.

Like satellite dishes, four-foot-wide ears catch sounds. On a hot day, an elephant flaps its ears to stay cool.

SOUTHERN ELEPHANT SEAL

Larger than their northern relatives, southern elephant seals feed on fish, squid, octopuses, and seabirds in southern waters. Males may be twice as long and four times as heavy as females.

Elephant seals, like these females, relax during the day and feed at night.

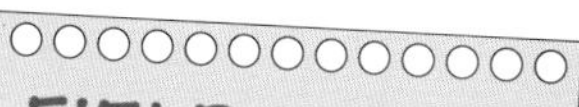

FIELD NOTES

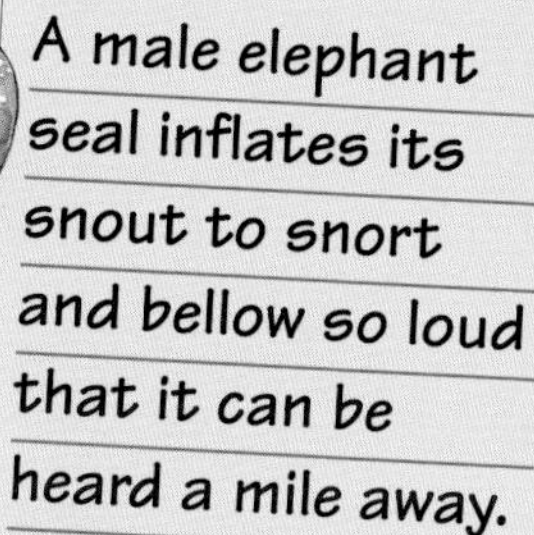

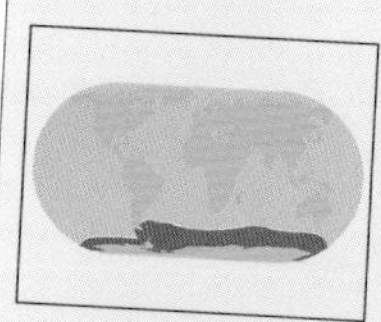

WHERE TO FIND:

Largest of all seals, southern elephant seals range northward from the waters around Antarctica.

WHAT TO LOOK FOR:

✱ SIZE
Elephant seals weigh as much as 4 tons and grow more than 20 feet long.

✱ COLOR
Males are dark gray to grayish brown. Females have browner bodies.

✱ BEHAVIOR
No mammal dives deeper—nearly a mile below the surface.

✱ MORE
This seal got its name from the male's trunk-like snout.

HIPPOPOTAMUS

Although its name means "river horse," the hippopotamus (hip-uh-POT-uh-muss) is only distantly related to the horse. Good swimmers, hippos stay in water much of the day.

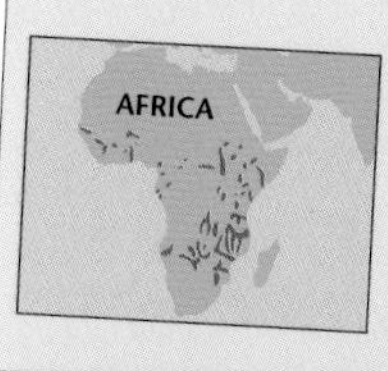

WHERE TO FIND:

The hippopotamus makes its home in and near rivers, lakes, and streams in central and southern Africa.

WHAT TO LOOK FOR:

✳ SIZE

A hippo weighs about 4 tons and reachs 15 feet in length.

✳ COLOR

It is dark brown to purplish gray.

✳ BEHAVIOR

At night, a hippo leaves the water and walks as far as five miles to eat grass.

✳ MORE

A hippo sleeps in water, rising every couple of minutes to breathe.

FIELD NOTES

Like a crocodile, a hippopotamus can rest with only its eyes, ears, and snout showing above the water.

When excited or threatened, a hippo opens its huge mouth and displays its sharp tusks.

WHITE RHINOCEROS

The white rhinoceros (rye-NOSS-uh-russ) is the largest of five different kinds of rhinos. Rhinos use their horns to protect themselves. The biggest horn ever measured was five feet long.

FIELD NOTES

While a rhinoceros grazes, oxpeckers get free rides and meals. They dine on insects clinging to the rhino's hide.

Although adult rhinos snort, snarl, and roar at each other, mothers mew softly to their calves.

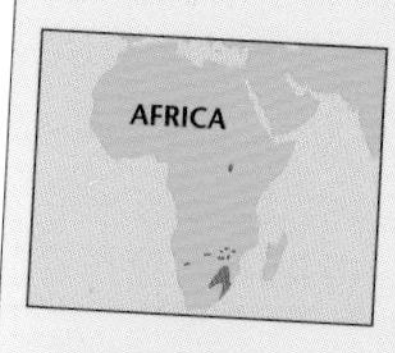

WHERE TO FIND:

The white rhinoceros lives in woodlands and on grasslands in central and southern Africa.

WHAT TO LOOK FOR:

✱ SIZE
Six feet tall at the shoulder, the largest rhinoceros weighs almost four tons.

✱ COLOR
The white rhino is actually gray.

✱ BEHAVIOR
Most of the time, rhinos calmly feed on grasses in groups of about a dozen animals.

✱ MORE
The rhino can run faster than a person.

WALRUS

The walrus is the only pinniped (PIN-ih-ped), or fin-footed sea mammal, that has long, down-thrusting tusks. These are handy for competing with rival walruses and fighting killer whales or polar bears.

FIELD NOTES

Like giant ice hooks, 40-inch-long tusks help the walrus haul its huge body out of the water.

The walrus's lumpy, two-inch-thick hide covers a layer of fat as much as six inches thick.

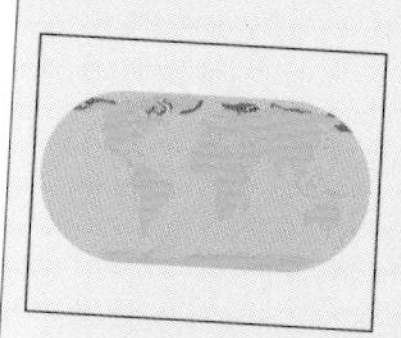

WHERE TO FIND:

The walrus swims in shallow waters of the Arctic Ocean and nearby seas. It comes ashore to rest.

WHAT TO LOOK FOR:

✱ SIZE
The largest walruses weigh almost two tons.

✱ COLOR
Walruses are brown or tan.

✱ BEHAVIOR
A walrus can sleep in water. To hold its head up, it inflates air sacs in its neck.

✱ MORE
A walrus eats about 100 pounds a day, devouring as many as 6,000 clams in one meal.

GIRAFFE

As tall as trees, giraffes tower over all other land mammals. A newborn is the size of a grown person and doubles its height in a year. Great height, vision, and speed help giraffes spot and outrun lions.

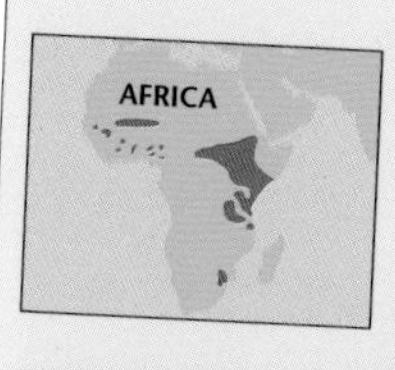

WHERE TO FIND:

Giraffes are found in Africa—on grasslands and plains and in woodlands—south of the Sahara.

WHAT TO LOOK FOR:

✱ SIZE
Giraffes grow to be 19 feet tall and weigh 1.5 tons.

✱ COLOR
Giraffes are tan with brown spots.

✱ BEHAVIOR
Just after sunrise, giraffes can be seen pulling leaves from treetops with their foot-and-a-half-long tongues.

✱ MORE
Giraffes nap standing up. At night, they sometimes lie down to sleep.

FIELD NOTES

To drink at a water hole, a giraffe spreads its six-foot-long legs apart and lowers its neck.

Giraffes wander alone or in small groups while feeding during the day.

NORTHERN SEA LION

Like lions of Africa, the sea lion roars. It walks on all fours, too—all four flippers, that is. The northern sea lion is the largest of five kinds of sea lions.

WHERE TO FIND:
The northern sea lion inhabits coastal waters and rocky shores from Japan to southern California.

WHAT TO LOOK FOR:

✱ SIZE
Sea lion bulls reach 11 feet in length and weigh more than a ton.

✱ COLOR
Northern sea lions range from yellowish brown to brown.

✱ BEHAVIOR
The sea lion can dive nearly 600 feet to find fish, squid, and octopuses.

✱ MORE
Sea lion pups bleat like lambs.

FIELD NOTES

The sea lion uses its flippers like feet. To "walk," it turns its hind flippers forward and lifts its body.

Gold, lion-like eyes and a bellowing roar inspired the sea lion's name.

AMERICAN BISON

Humped shoulders and a massive head with short, curved horns make the bison easy to identify. North America's heaviest land animal, the bison is sometimes called the American buffalo.

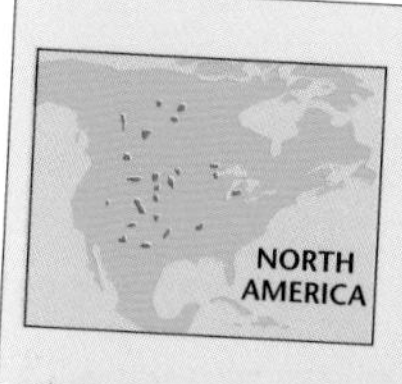

WHERE TO FIND:

Bison roam grasslands and open woodlands in several North American parks and refuges.

WHAT TO LOOK FOR:

✳ SIZE

A bison can weigh 2,200 pounds and stand 6 feet tall at the shoulder.

✳ COLOR

The bison is light to dark brown.

✳ BEHAVIOR

To get rid of insects and relieve itching, a bison takes a dust bath. It rolls back and forth in a shallow depression called a wallow.

✳ MORE

Bison eat mostly grass and some berries.

Short, pointed horns and a thick mat of dark brown hair crown the bison's head. Males and females have horns.

FIELD NOTES

A herd galloping from danger sounds like rolling thunder. Bison can run at speeds up to 30 mph.

GAUR

The world's largest wild cattle have big blue eyes. Gaur—the name rhymes with "power"—live in herds of up to 40 animals. For much of the day, they quietly graze on grasses and munch on leaves.

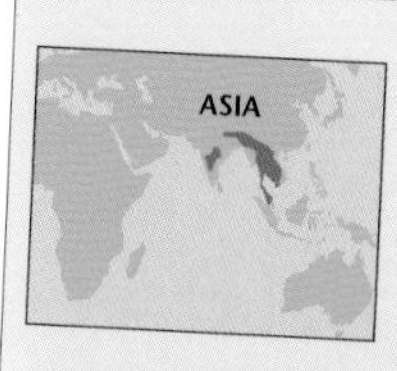

WHERE TO FIND:
Gaur can be found in the forested hills of India and Southeast Asia, usually near water or in grassy areas.

WHAT TO LOOK FOR:

✱ SIZE
A gaur may weigh more than a ton and stand six feet tall at the shoulder.

✱ COLOR
The gaur is dark brown with light brown or white markings on its legs.

✱ BEHAVIOR
Gaur usually move about during the day. They are shy animals, and become nocturnal if humans live nearby.

✱ MORE
Gaur roam in herds led by one male.

FIELD NOTES

Threatened by a tiger lurking in the grass, a gaur tries to look as big as possible by turning sideways.

A helpless gaur calf relies on its mother to protect it from harm.

WEST INDIAN MANATEE

As big as cows, manatees are plant-eaters, or herbivores (UR-buh-vorz), too. Manatees graze on plants underwater. The West Indian manatee is the largest of three kinds of manatees.

FIELD NOTES

Manatees like to nuzzle and play. Snout to snout, they seem to greet each other with a kiss.

Moving slowly through shallow water, a manatee places one flipper in front of the other to walk on the bottom of a river.

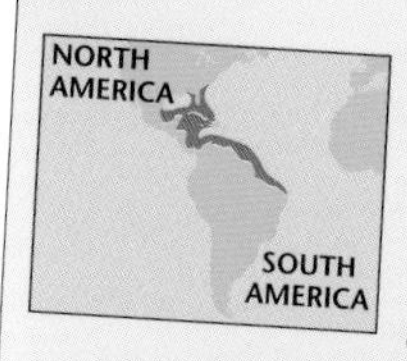

WHERE TO FIND:

The West Indian manatee can be found in warm coastal waters from the southern U.S. to Brazil.

WHAT TO LOOK FOR:

✷ SIZE
The West Indian manatee weighs as much as a ton and is 15 feet long.

✷ COLOR
Manatees are gray.

✷ BEHAVIOR
Manatees feed on aquatic plants. They are the only plant-eating mammals that live entirely in water.

✷ MORE
A manatee can close its nostrils to keep out water.

ELAND

Largest antelope in the world, the eland (EE-lund) frequently astounds onlookers with its high-jumping ability. From a standing position, a young eland can easily leap over a nine-foot fence.

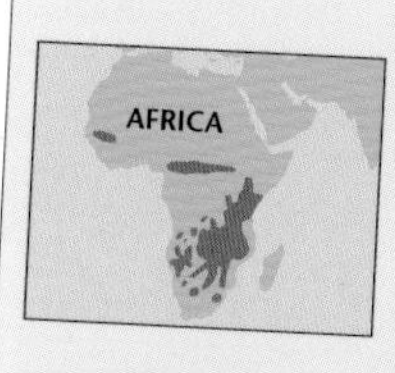

WHERE TO FIND:

Herds of elands wander among wooded and grassy areas of Africa south of the Sahara.

WHAT TO LOOK FOR:

✱ **SIZE**
Elands stand almost six feet tall at the shoulder and weigh about one ton.

✱ **COLOR**
Elands are tan to bluish gray.

✱ **BEHAVIOR**
These animals eat leaves, grass, fruits, and seeds. They may use their horns to pull down branches or plow up plants.

✱ **MORE**
Young elands seem to be more attached to each other than to their mothers.

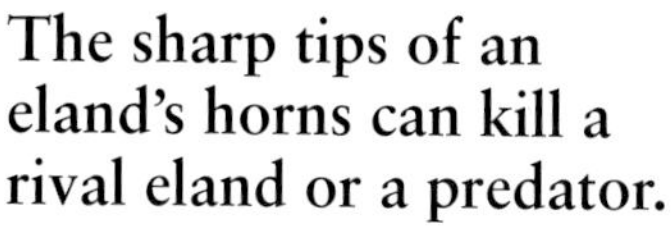

The sharp tips of an eland's horns can kill a rival eland or a predator.

BACTRIAN CAMEL

Known as "ships of the desert," camels cross vast dry lands. When food and water are scarce, camels survive on fat stored in their humps. Bactrian camels can weigh up to 200 pounds more than Arabian camels.

FIELD NOTES

The shaggy Bactrian (BACK-tree-un) camel has two humps; the sleek Arabian camel has one.

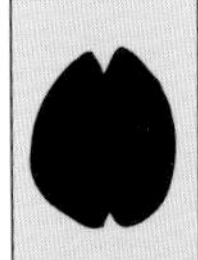

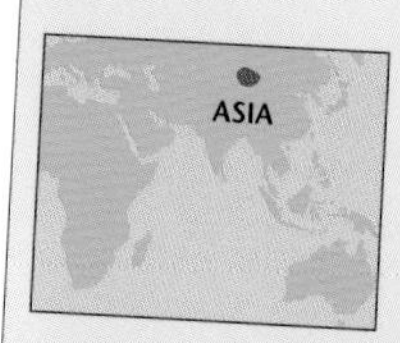

WHERE TO FIND:

Named for Bactria, an ancient country, this camel is found in rocky deserts of central Asia.

WHAT TO LOOK FOR:

✱ SIZE
Seven feet tall at the hump, Bactrian camels can weigh 1,800 pounds.

✱ COLOR
Camels are light to dark brown.

✱ BEHAVIOR
All camels can get the water they need by nibbling desert plants.

✱ MORE
Because they don't sweat much, camels lose body moisture very slowly.

Two rows of eyelashes, hair-lined ears, and nostrils that close protect a camel from blowing sand.

MOOSE

Antlers like bony hat racks grow from the head of the moose—largest member of the deer family. The biggest moose of all lives in North America. From tip to tip, its antlers span six feet.

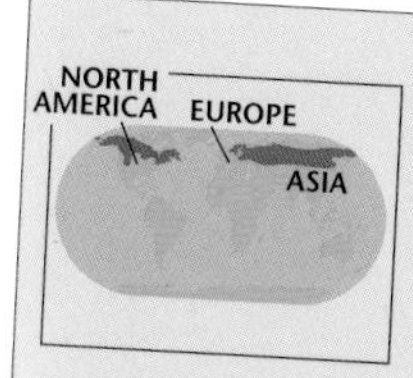

WHERE TO FIND:

The moose usually can be found near water in the northern forests of Europe, Asia, and North America.

WHAT TO LOOK FOR:

✷ SIZE
As tall as 7 feet at the shoulder, a moose weighs up to 1,800 pounds.

✷ COLOR
The moose is light to dark brown.

✷ BEHAVIOR
The moose lives alone in summer, but when winter snows fall, it joins other moose to form herds.

✷ MORE
Long, slender legs help the moose run faster than 30 miles an hour.

FIELD NOTES

Moose like to eat water plants. They will swim as long as 2 hours and dive as deep as 18 feet to feed.

Soft skin called velvet covers and protects new antlers.

YAK

On mountain slopes almost four miles above sea level lives a great ox known as the yak. Strong and surefooted, it easily climbs peaks where snow never melts. Few great mammals live at such heights.

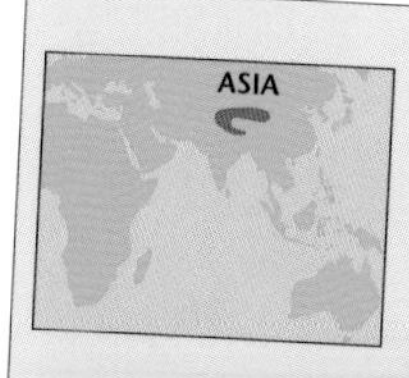

WHERE TO FIND:

Wild yaks live only in central Asia, roaming windswept highlands and steep mountain ranges.

WHAT TO LOOK FOR:

✳ SIZE
A yak weighs as much as 1,800 pounds and measures 6 feet at the shoulder.

✳ COLOR
Yaks have thick brown or black coats.

✳ BEHAVIOR
Always on the lookout for wolves, female yaks and calves travel in large herds. Males form smaller groups.

✳ MORE
If threatened, yaks stand together, lower their heads, and display their horns.

FIELD NOTES

At home in the snow, yaks use their muzzles and hooves to uncover grasses. They also eat snow.

Its thick, shaggy coat keeps the yak warm in its highland home.

BROWN BEAR

Brown bears are the biggest carnivores (CAR-nuh-vorz), or meat-eaters, that live on land. The largest brown bears live in Alaska. In summer and fall they eat huge meals, gaining as much as 400 pounds.

FIELD NOTES

To catch salmon, a brown bear belly flops into the water and grabs the fish with its paws or jaws.

Big and bulky—but quick—a brown bear can sprint as fast as a horse to nab small prey.

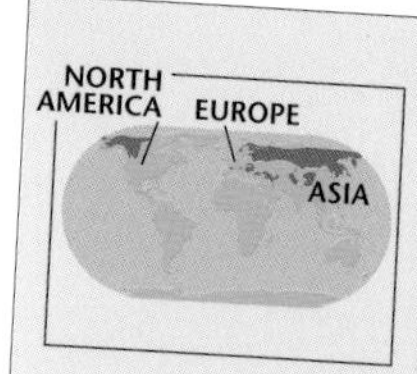

WHERE TO FIND:

Brown bears live in a variety of habitats in Europe, Asia, and North America.

WHAT TO LOOK FOR:

✷ **SIZE**
The largest brown bears weigh 1,700 pounds and are 10 feet long.

✷ **COLOR**
Brown bears range in color from tan to nearly black.

✷ **BEHAVIOR**
They eat small mammals, fish, insects, berries, leaves, twigs, and grasses.

✷ **MORE**
Bears make dens and sleep through the coldest part of winter.

POLAR BEAR

The great bear of the far north—the polar bear—swims well. Air-filled hairs help keep it afloat. Fur on the feet of polar bears helps them stay warm and keeps them from slipping on icy surfaces.

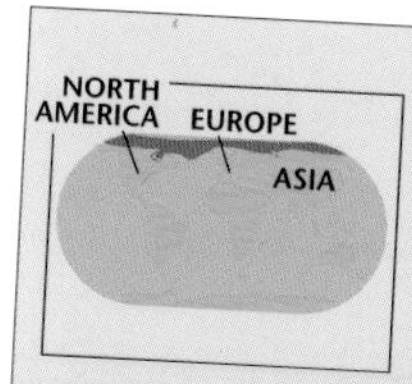

WHERE TO FIND:

The polar bear lives on Arctic sea ice and the northernmost coasts of Europe, Asia, and North America.

WHAT TO LOOK FOR:

✱ SIZE
A polar bear can weigh almost 1,700 pounds and grow to 9 feet long.

✱ COLOR
Its coat is white to yellowish white.

✱ BEHAVIOR
Polar bears eat seals, walruses, small mammals, birds, fish, shellfish, and some plants.

✱ MORE
Unlike other bears, polar bears can be active throughout the year.

A polar bear's sensitive nose locates prey hiding in snowdrifts.

WATER BUFFALO

Largest wild buffalo, the water buffalo spends much of its time submerged in water. Only its head rises above the surface. Seeking safety in numbers, water buffalo usually live in large herds.

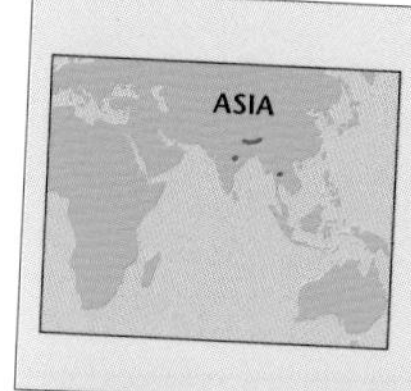

WHERE TO FIND:

Water buffalo can be found along rivers in southern and southeastern Asia.

WHAT TO LOOK FOR:

✱ SIZE
Water buffalo weigh 1,500 pounds and stand 5 feet tall.

✱ COLOR
They are gray or black.

✱ BEHAVIOR
During the day, water buffalo stand in water to keep their nearly hairless skin from drying out.

✱ MORE
People have tamed water buffalo and use them to pull carts and carry loads.

FIELD NOTES

To protect its sensitive skin from the bites of insects, the water buffalo rolls its huge body in mud.

Long, curving horns measure up to six feet from tip to tip.

AMERICAN ELK

This member of the deer family has tall, branching antlers that may reach five feet from base to tip. An old bull may sport a pair of antlers with 15 sharp points. Smaller elk live in central Asia.

FIELD NOTES

Each year elk grow new antlers. Velvety skin covers them in the summer and peels off in the fall.

Great racks of antlers crown the brows of two bull elk. Rivals clash their antlers in mating contests.

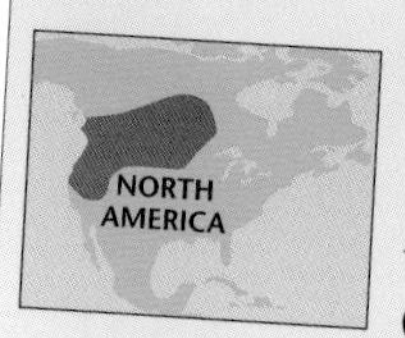

WHERE TO FIND:

The American elk lives in mountain pastures and woodlands of western Canada and the U.S.

WHAT TO LOOK FOR:

✱ SIZE
The American elk can weigh 1,100 pounds and stand 5 feet tall at the shoulder.

✱ COLOR
The elk is light to dark brown.

✱ BEHAVIOR
Elk feed on grasses, herbs, and tree bark.

✱ MORE
A bull elk makes a trumpeting sound known as a bugle.

MUSK OX

Musk oxen are distantly related to goats and antelopes. Sharp rims and soft pads on their hoofs help musk oxen climb and paw through snow to find food. Two layers of hair keep the animal warm.

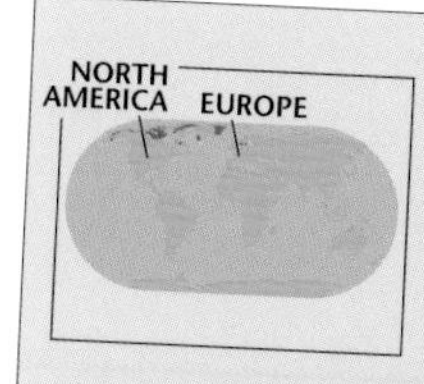

WHERE TO FIND:

Herds of musk oxen can be found on the Arctic tundra of North America and Scandinavia.

WHAT TO LOOK FOR:

✷ SIZE
A musk ox weighs as much as 900 pounds and stands 5 feet tall.

✷ COLOR
The animal's coat is dark to light brown, and its legs are nearly white.

✷ BEHAVIOR
Using its horns, a musk ox can throw a wolf into the air. The musk ox's hoofs crush the wolf when it lands.

✷ MORE
Musk oxen eat grasses and willows.

FIELD NOTES

Facing toward a threat, musk oxen form a protective circle around their calves.

A shaggy musk ox sheds its woolly undercoat when summer comes.

MALAYAN TAPIR

At night, a shy creature called a tapir comes out of hiding. It looks like a pig but is related to the horse. The Malayan tapir, largest of several kinds of tapirs, has markings that blend with forest shadows.

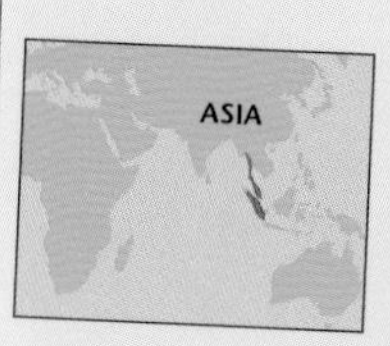

WHERE TO FIND:

The Malayan tapir lives in Myanmar, Thailand, Malaysia, and on the island of Sumatra.

WHAT TO LOOK FOR:

✳ SIZE
The Malayan tapir is almost 4 feet tall at the shoulder and weighs 800 pounds.

✳ COLOR
Also called the blanket tapir, it is black with a large white patch on its back.

✳ BEHAVIOR
To get rid of pesky insects, tapirs scratch their chests with their hind feet like dogs.

✳ MORE
Tapirs are good swimmers.

Watermelon-like stripes fade, and a white patch like a blanket appears as the Malayan tapir grows older.

FIELD NOTES

A tapir's snout forms a short trunk, which it uses to grab leaves and pull them to its mouth.

WILD BOAR

World's largest wild hog, the wild boar eats almost anything. And it fears almost nothing. To frighten away intruders, it bristles, grunts, then charges with its razor-sharp tusks ready to gore an enemy.

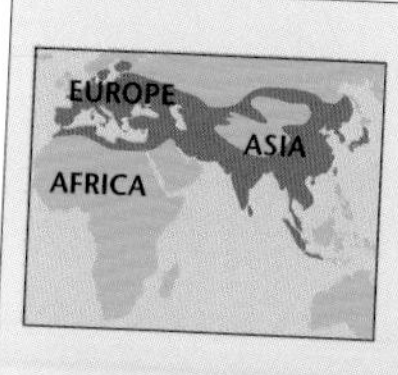

WHERE TO FIND:

The wild boar lives in many parts of Europe, Asia, and northern Africa. It prefers wooded, scrubby areas.

WHAT TO LOOK FOR:

✱ SIZE

The wild boar grows more than 3 feet tall at the shoulder and weighs 700 pounds.

✱ COLOR

The boar is light to dark brown.

✱ BEHAVIOR

To mark a territory and show other boars how tall it is, a male marks trees as high as it can reach with its tusks.

✱ MORE

Boars build nests of leaves.

On the alert, a wild boar stiffens, cocks its ears, and sniffs the air.

FIELD NOTES

A boar sniffs for a truffle, a food that grows underground, then digs it up with its long, strong nose.

TIGER

Largest member of the cat family, a tiger can eat 50 pounds of meat in one night. It hunts animals two or three times its size. After stalking its prey, a tiger leaps as far as 30 feet to bring it down.

How do you tell tigers apart? Look at the stripes: No two tigers have the same pattern.

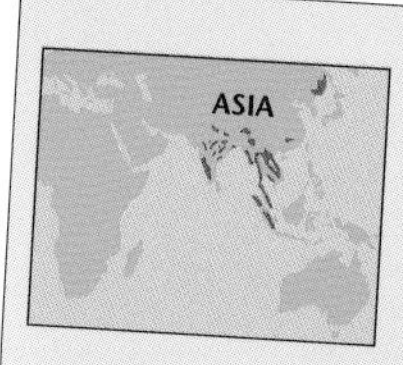

WHERE TO FIND:

Tigers can be found only in a few areas of India, eastern Russia, and southeastern Asia.

WHAT TO LOOK FOR:

✷ SIZE
Tigers weigh up to 675 pounds and are more than 6 feet long.

✷ COLOR
They are brown to reddish orange, with dark stripes and white undersides.

✷ BEHAVIOR
Tigers try to remain invisible, hiding in forests or lurking at forest edges.

✷ MORE
Except for females with cubs, tigers live by themselves.

OKAPI

Scientists did not even know the okapi (oh-COP-ee), a relative of the giraffe, existed until a hundred years ago. Larger than a lion, the shy okapi stays hidden by quietly roaming dark rain forests.

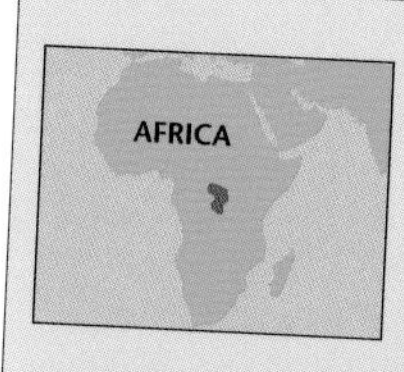

WHERE TO FIND:

The okapi can be found wandering through forested river valleys in a small area of central Africa.

WHAT TO LOOK FOR:

✱ SIZE
Weighing 550 pounds, the okapi stands almost six feet tall at the shoulder.

✱ COLOR
It is dark brown with stripes.

✱ BEHAVIOR
To protect her newborn calf from predators, a female hides it. The two cough and whistle to communicate.

✱ MORE
Okapis eat leaves of trees and shrubs.

An okapi has zebra-like stripes on its rump. A male has knobs like a giraffe's on his forehead.

BLUE WILDEBEEST

The wildebeest (WILL-duh-beast), or gnu (NYOO), is a kind of antelope. It seems to have an antelope's legs, an ox's head, a horse's tail and mane, and a bison's horns. There are two kinds of wildebeests. The larger has blue skin.

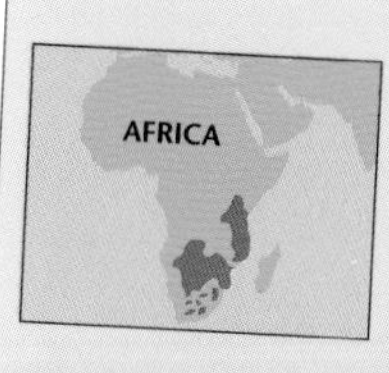

WHERE TO FIND:

The blue wildebeest lives on grasslands in eastern and southeastern Africa.

WHAT TO LOOK FOR:

✱ SIZE
A blue wildebeest weighs 550 pounds and is nearly 5 feet tall at the shoulder.

✱ COLOR
It is grayish blue.

✱ BEHAVIOR
When males stake out territories, they snort and grunt, leap about frantically, and shake their heads.

✱ MORE
Huge herds of wildebeests migrate from summer to winter pastures.

Blue gnus drink from a water hole. A herd of wildebeests stays in an area till the water runs out. Then it moves on.

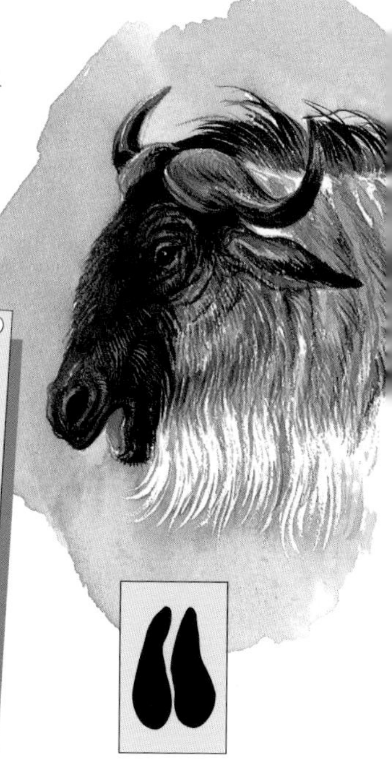

FIELD NOTES

Some blue wildebeests grow white beards. Wildebeest bulls grunt like big frogs.

LION

Called the King of Beasts, the lion is Africa's largest carnivore. A lion is so strong that it can bring down prey twice its size. Males grow great manes that make their heads look enormous.

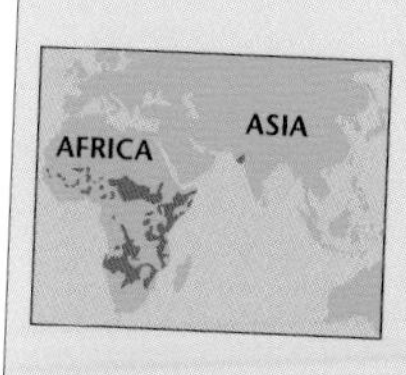

WHERE TO FIND:
Lions live in Africa's grasslands, plains, and open woodlands. A small group remains in India.

WHAT TO LOOK FOR:

✱ **SIZE**
The lion weighs more than 500 pounds and is 6 feet long.

✱ **COLOR**
Lions range from light to dark brown.

✱ **BEHAVIOR**
By roaring, a lion lets other lions know where it is. A roar can be heard five miles away.

✱ **MORE**
Lions kill alone or in groups. Males almost always feed before females.

Surveying his kingdom, a lion relaxes in the grass. Lions rest as much as 20 hours a day.

FIELD NOTES

Lions live together in groups called prides. As many as 40 lions may make up a pride.

GORILLA

Greatest of the apes, a male gorilla (guh-RIL-uh) stands and thumps its chest to impress females or frighten rivals. Gorillas spend much time feeding, grooming each other, and playing with their young.

FIELD NOTES

Gorillas are very gentle. Young ones receive tender care and stay close to mother, often riding on her back.

When a male gorilla grows a patch of white hair as it ages, it is called a silverback.

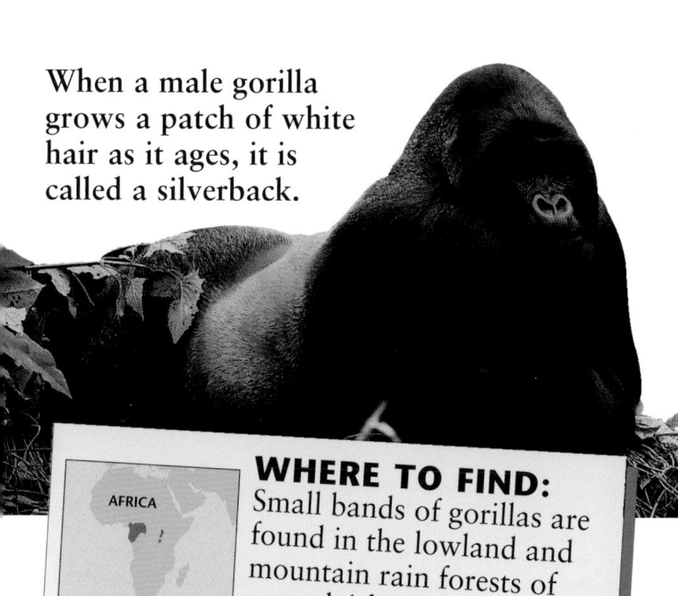

WHERE TO FIND:

Small bands of gorillas are found in the lowland and mountain rain forests of central Africa.

WHAT TO LOOK FOR:

✱ SIZE
A gorilla can weigh 450 pounds and grow 6 feet long.

✱ COLOR
It has blue-black to brownish gray hair.

✱ BEHAVIOR
Gorillas live in groups of 2 to 20 animals and feed on roots, tree bark, bamboo shoots, and wild celery.

✱ MORE
Gorillas spend nearly half the day resting in nests made of leafy plants.

GIANT PANDA

One of the rarest large mammals, the giant panda is as big as a barrel. Pandas munch bamboo leaves, stems, and shoots. In one day, a giant panda may eat as much as 20 pounds of bamboo.

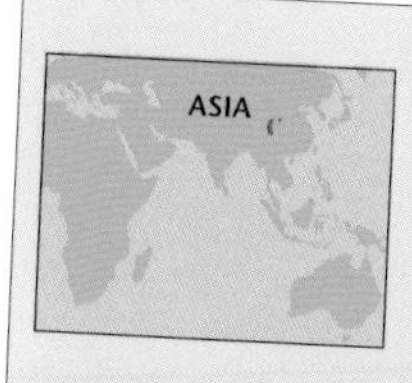

WHERE TO FIND:

The giant panda can be found in a few bamboo forests on mountain slopes of central China.

WHAT TO LOOK FOR:

✱ SIZE
Weighing as much as 300 pounds, the giant panda grows 5 feet long.

✱ COLOR
Pandas have distinctive white and black markings.

✱ BEHAVIOR
Besides bamboo, pandas eat flowers, birds, and small mammals.

✱ MORE
A mother panda may be 800 times heavier than her newborn cub.

FIELD NOTES

Five toes on each forepaw and long, thumb-like wrist bones make grasping bamboo easy for a panda.

Feeding on bamboo, a panda breaks the stem and crunches it into shreds.

SIBERIAN IBEX

You can tell the age of an ibex (EYE-beks), a kind of wild goat, by counting the ridges on its horns. The Siberian ibex is the largest of several types of ibexes and has the biggest horns.

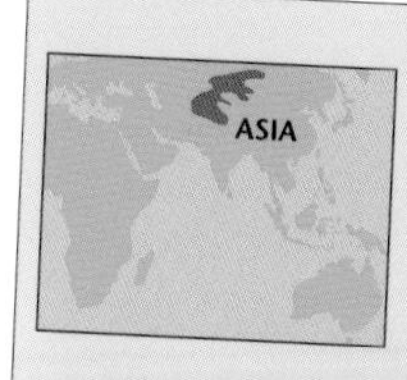

WHERE TO FIND:
The Siberian ibex lives in the high mountains of central Asia.

WHAT TO LOOK FOR:

✱ SIZE
The Siberian ibex is 3.5 feet tall at the shoulder and weighs 275 pounds.

✱ COLOR
It is brownish gray.

✱ BEHAVIOR
The ibex uses its forelegs to scrape away snow and find food in its high mountain home.

✱ MORE
Smaller ibexes are found in Europe and Africa.

Female as well as male ibexes grow horns. Those of the female are shorter.

RED KANGAROO

Able to leap a six-foot-tall man in a single bound, male red kangaroos are the largest roos. The females, however, are the fastest. They are called blue fliers because they have bluish fur.

FIELD NOTES

A baby kangaroo, called a joey, rides in a pouch on its mother's belly for about six months.

There are 23 kinds of kangaroos, ranging in size from male red roos like this one to rat-size creatures.

WHERE TO FIND:

Large numbers of red kangaroos range across the great open plains of central and western Australia.

WHAT TO LOOK FOR:

✱ SIZE
The red kangaroo stands as tall as 6 feet and weighs 150 pounds.

✱ COLOR
Males are light to reddish brown. Females are bluish gray.

✱ BEHAVIOR
Males fight by balancing on their big tails and kicking with their hind feet.

✱ MORE
In just one jump, a red kangaroo can leap as far as 25 feet.

GRAY WOLF

Largest wild member of the dog family, the gray wolf roams in small groups called packs. Wolves sometimes claim hunting areas 60 miles wide. They work together to bring down big game.

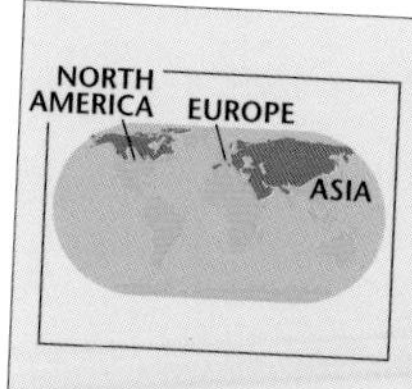

WHERE TO FIND:

The gray wolf lives in widely scattered populations across North America, Europe, and Asia.

WHAT TO LOOK FOR:

✱ SIZE
Including its tail, the gray wolf measures more than six feet long. It weighs up to 120 pounds.

✱ COLOR
It can be gray, white, or nearly black.

✱ BEHAVIOR
Gray wolves hunt in packs, usually at night but sometimes during the day.

✱ MORE
To communicate, gray wolves growl and howl, whimper and whine.

Thick fur keeps a wolf warm even at minus 40°F.

FIELD NOTES

The way a wolf carries its tail shows its rank. A pack leader holds its tail high; followers lower theirs.

CAPYBARA

Related to the rat, the capybara (cap-ih-BAR-uh) is the world's largest rodent. As many as 20 capybaras may roam together as a herd. When threatened, they plunge into water and quickly paddle away.

FIELD NOTES

The capybara's front teeth never stop growing. Chewing tough plants helps keep them worn down.

Wading in, a capybara looks around for tasty water plants.

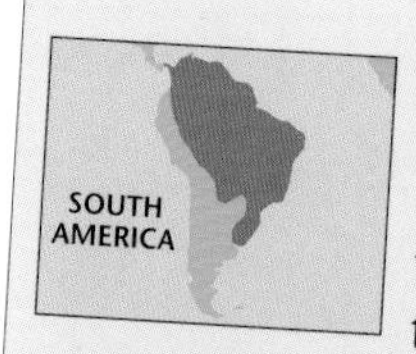

WHERE TO FIND:

The capybara ranges across well-watered areas east of the Andes, from Panama to northern Argentina.

WHAT TO LOOK FOR:

✷ SIZE
The capybara is more than 4 feet long and weighs about 110 pounds.

✷ COLOR
Its hair is light to dark brown.

✷ BEHAVIOR
To rest or sleep, the capybara digs a shallow bed in the ground.

✷ MORE
Because its hair is thin, the capybara keeps its exposed skin from drying out by wallowing in mud or water.

GIANT ANTEATER

Even big cats fear the giant anteater when it stands on its hind feet and strikes with its four-inch claws. The giant anteater is by far the largest of several kinds of anteaters.

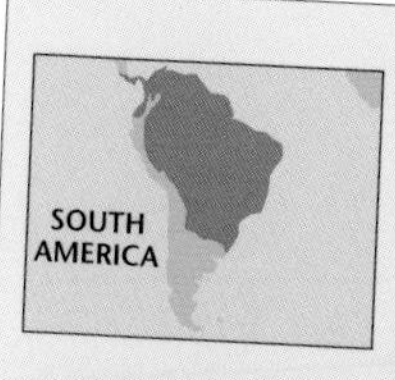

WHERE TO FIND:

The giant anteater lives in swamps and grasslands of Central and South America.

WHAT TO LOOK FOR:

✳ SIZE
Weighing nearly 90 pounds, the giant anteater is 7 feet long from head to tail.

✳ COLOR
Its coat is gray with a dark stripe.

✳ BEHAVIOR
The anteater spends most of the day looking for food. It eats insects, especially ants and termites, and fruit.

✳ MORE
On cold nights, the giant anteater uses its bushy, 3-foot-long tail as a blanket.

FIELD NOTES

An anteater rips open a mound, pokes its nose in, and collects termites on its sticky tongue.

Clinging to its mother's back so that their stripes line up, a young anteater may not be seen by predators.

FLYING FOX

The group of largest bats gets its name from the fox-like head these mammals have. Of 59 different kinds of flying foxes, the largest has wings that stretch six feet from end to end.

FIELD NOTES

The wingspan of the largest flying fox is about as great as that of a bald eagle.

Wings tucked away for the day, the flying fox hangs by its feet in a tree. Hundreds of these bats may share a single roosting place.

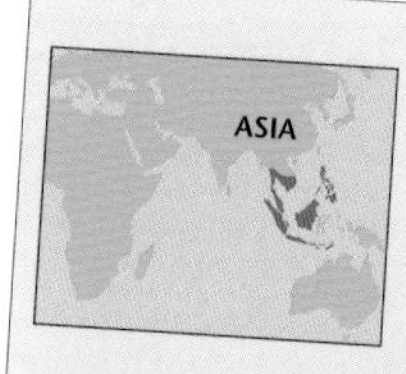

WHERE TO FIND:
The largest flying fox lives in forests and swamps of the Philippines, Thailand, Malaysia, and Indonesia.

WHAT TO LOOK FOR:

✷ SIZE
The largest flying fox weighs 2 pounds and has a 16-inch-long body.

✷ COLOR
It is grayish brown or black, with a light area on the shoulders.

✷ BEHAVIOR
This bat feeds and rests in a fruit tree. It bites into fruit, swallows the juice, and spits out the seeds.

✷ MORE
It may even drink seawater.

GLOSSARY

air sac An internal pouch that fills with air.

antlers The pair of bony growths on the heads of the males of most deer species. Deer replace their antlers each year. The old pair falls off and a new pair grows in.

bull The adult male of some large animal species.

carnivore An animal that eats meat. Its sharp teeth can cut and tear flesh.

dolphin A member of a family of small toothed whales. Dolphins are sometimes called porpoises. They have cone-shaped teeth.

family A group of animals having common characteristics.

graze To feed on grasses.

herbivore An animal that eats plants.

horns Hard, hair-like material surrounding bony cores on the heads of hoofed mammals. Unlike antlers, horns are permanent.

migrate To move from one place to another when one season ends and another begins.

nocturnal Active at night.

pinniped A "fin-footed" mammal that has flippers and spends much of its life in water. It gives birth on land.

points Tips of a deer's antlers.

predator An animal that hunts and eats other animals.

prey An animal hunted for food.

rack A pair of antlers.

rodent A mammal with chisel-like front teeth that grow throughout the animal's life.

territory A place occupied and defended by an animal or several related animals from others of the same kind.

truffle An edible fungus that grows underground.

tundra A flat, treeless plain in Arctic and subarctic regions.

tusks Large teeth that stick out when an animal's mouth is closed.

velvet Soft skin that covers and nourishes the growing antlers of a deer.

INDEX OF GREAT MAMMALS

ABOUT THE CONSULTANT

Ronald M. Nowak worked as a mammalogist in the endangered species program of the U.S. Fish and Wildlife Service for 24 years. Dr. Nowak is the author of the fourth and fifth editions of *Walker's Mammals of the World*, and he is currently preparing the sixth edition. He has published some 70 papers, articles, and books of scientific and popular interest.

PHOTOGRAPHIC CREDITS

Photographs supplied by Animals Animals/Earth Scenes.

front cover Gérard Lacz **back cover** Johnny Johnson **1** Johnny Johnson **3** Johnny Johnson **4-5** Gérard Lacz **7** Doc White/Images Unlimited **9** James D. Watt **11** Johnny Johnson **12** Lynn M. Stone **15** Gérard Lacz **17** Gérard Lacz **19** Ted Levin **21** Peter Weimann **23** Shane Moore **25** Charles Palek **27** Mickey Gibson **29** Steven David Miller **31** Zig Leszczynski **33** Robert Maier **35** Johnny Johnson **37** Peter Weimann **39** Leo Keeler **40** Johnny Johnson **43** Phil Degginger **45** Charlie Palek **47** Robert Maier **49** Gérard Lacz **51** Robert Maier **53** Zig Leszczynski **55** Zig Leszczynski **57** Ana Laura Gonzalez **59** Leonard Lee Rue, III **61** B.G. Murray, Jr. **63** Lynn M. Stone **65** Richard Kolar **67** Prenzel Photo **69** Peter Weimann **71** Joe McDonald **73** John Chellman **75** Bruce Watkins **76** Gérard Lacz **79** Ken Cole.